Contents

Introduction	4
Index of visual symptoms by location on the tree	5
Pests and diseases	6
Suggested management programme for the control of pests and diseases	76
Table 1a: Plant protection products registered for use in an amenity environment – Insecticides	77
Table 1b: Plant protection products registered for use in an amenity environment – Fungicides	78
Reporting pests and diseases	79
Biosecurity	80
Index of scientific names	82
Index of common names	83

Acknowledgements

The Arboricultural Association gratefully acknowledges the support and technical advice provided by The Forestry Commission, Forest Research and Bartlett Tree Experts, Ltd, who have supplied images and technical editing during the development of this publication.

Forest Research is part of the Forestry Commission. It carries out world-class scientific research and technological development relevant to forestry for a range of internal and external clients.

For further information visit:

Forestry Commission: www.forestry.gov.uk

Forest Research: www.forestry.gov.uk/forestresearch.

Bartlett Tree Expert: www.bartlett.com

Our thanks also to David Lonsdale for his support and editorial expertise and to Sarah Kiss for proof checking and attention to detail.

Introduction

Arborists commonly see a range of pests, diseases and disorders that affect amenity trees in the UK. This book is designed to help identify those commonly encountered, in a simple, easy-to-use format.

Of course, all trees are natural hosts to a range of other organisms, and many of these are not only harmless but actually beneficial to the good health of the tree and its environment. Even those that may be perceived to be causing harm to the tree may also be essential to the ecosystem of which the tree is a part. Surrounding trees, shrubs and plants, local fauna, the soil and all the organisms that live in it are reliant on each other to some degree. Therefore identifying what is present and what is causing damage is an essential part of evaluating whether any external action is required.

Pests are normally considered to be living organisms that cause harm to another either through direct damage or by carrying in a disease, when they may also be described as a vector of a pathogen. A disease will cause specific signs or symptoms in the host that are detrimental to normal function. For example, *Neonectria coccinea* (a fungal pathogen) can induce beech bark disease but only if the host tree is physiologically stressed. It may be that the pest *Cryptococcus fagisuga* (the felted beech scale insect) adds to that stress, allowing the disease to colonise the tree, but the pest does not itself vector the disease. Dutch elm disease is caused by the pathogen *Ophiostoma novo-ulmi*, a fungus, and is spread by the pest *Scolytus*, the elm bark beetle.

A diagnosis of the pathogens affecting a tree will mostly result from observation of the symptoms present. Follow a methodical process to examine all aspects carefully so that no important detail is overlooked. Do not jump to conclusions but continue with your process until it is complete. If anything does not fit, repeat the process to ensure a correct result.

Symptoms will give an initial starting point for diagnosis but will not provide the full answer. For example, leaf damage (the symptom) may have a number of causes: fungal or bacterial infection, drought, insect activity or response to adverse site conditions. The symptoms will, however, be your starting point and it is from them that your investigation will start and expand. Then the causal agent will be determined.

A general diagnostic procedure could work as follows:

1. Preliminary assessment of the tree – to identify the species; assess the general condition, age and location; assess its relationship to other trees and any physical site conditions that may be relevant;
2. Symptoms – to provide a description, location or distribution and stage of development and thereby to locate the infection court;
3. Comparison – to other trees of the same species in the locality to determine spread and severity within a population;
4. Detailed assessment – of the infection court to identify the causal agent;
5. Diagnosis – complete with a review to test reliability against known information.

Index of visual symptoms by location on the tree

Symptom location	
General overall decline	*Armillaria* (honey fungus)
	Oak decline
	Ophiostoma novo-ulmi (Dutch elm disease)
Roots and buttress	*Armillaria* (honey fungus)
	Phytophthora (root rot)
	Sesia apiformis (hornet moth)
Main stem	*Anoplophora glabripennis* (Asian longhorn beetle)
	Canker
	Cossus cossus (goat moth)
	Cryptococcus fagisuga (felted beech scale)
	Cryptostroma corticale (sooty bark disease)
	Neonectria coccinea (beech bark disease)
	Pseudomonas syringae pv. *aesculi* (bleeding canker of horse chestnut)
	Pulvinaria regalis (horse chestnut scale)
Principal branches	*Anoplophora glabripennis* (Asian longhorn beetle)
	Canker
	Phytophthora cactorum/citricola
	Pseudomonas syringae pv. *aesculi* (bleeding canker of horse chestnut)
	Pulvinaria regalis (horse chestnut Scale)
	Splanchnonema platani (Massaria canker)
Leaf and shoot	*Apiognomonia veneta* (anthracnose of London plane)
	Apiognomonia erythrostoma (cherry leaf scorch)
	Blumeriella jaapii (cherry leaf spot)
	Cameraria ohridella (horse chestnut leaf miner)
	Chalara fraxinea (ash dieback)
	Cinara cupressi (cypress aphid)
	Cristulariella depraedans (a leaf disease of maples) see *Rhytisma acerinum* (Tar spot of sycamore)
	Dendrolimus pini (pine tree lappet moth)
	Dothistroma septosporum (red band needle blight)
	Drepanopeziza sphaeroides (anthracnose of willow)
	Erwinia amylovora (fireblight)
	Euproctis chrysorrhea (brown tail moth)
	Guignardia aesculi (horse chestnut leaf blotch)
	Lymantria dispar dispar (European gypsy moth)
	Melampsora or *Melampsoridum* spp. (rusts of poplar, birch and willow)
	Monilinia laxa (blossom wilt and wither tip)
	Ophiostoma novo-ulmi (Dutch elm disease)
	Phytophthora ilicis (on holly)
	Rhytisma acerinum (tar spot of sycamore)
	Seridium cardinal (Coryneum canker)
	Leaf spots and blotches
	Taphrina spp. (witches' broom)
	Thaumetopoea processionea (oak processionary moth)
	Venturia spp. (scab)
	Zeuzera pyrina (leopard moth)

Anoplophora glabripennis

Asian longhorn beetle

An insect pest mainly from China and neighbouring countries which may be imported living in the larval stage in packaging wood.

Asian longhorn beetle will attack a wide range of common broadleaves although not generally highly lignified trees.

The adult is a large, obvious black-and-white beetle, up to 40mm long with very long black-and-white antennae. Whilst they normally stay close to their hatching ground, they can fly up to 2km. Adults mate and the female lays eggs on a scraped area of bark. When the eggs hatch the larvae burrow into the wood and commence maturation feeding. This may take up to 2 years and is contained within the host tree. Adults emerge around June from large circular holes, 10mm diameter, in the stem or main branch structure.

Distribution: Currently only established in a small area of Kent and subject to eradication measures.

Species affected: Known hosts include: *Acer* (maples and sycamores); *Aesculus* (horse chestnut); *Albizia* (mimosa, silk tree); *Alnus* (alder); *Betula* (birch); *Carpinus* (hornbeam); *Cercidiphyllum japonicum* (Katsura tree); *Corylus* (hazel); *Fagus* (beech); *Fraxinus* (ash); *Koelreuteria paniculata*; *Platanus* (plane); *Populus* (poplar); *Prunus* (cherry, plum); *Robinia pseudoacacia* (false acacia/black locust); *Salix* (willow, sallow); *Sophora* (=*Styphlonobium japonicum*) (pagoda tree); *Sorbus* (rowan, whitebeam etc.); *Quercus palustris* (American pin oak); *Quercus rubra* (North American red oak); and *Ulmus* (elm), but most trees may be capable of supporting this pest.

Treatment: All infected trees have to be destroyed on confirmation.

Note: This is notifiable pest and any suspected sightings of adults or emergence holes must be reported to the relevant authority. See page 79.

Apiognomonia/Blumeriella/Monilinia

Leaf and shoot diseases of cherry

Apiognomonia erythrostoma - cherry leaf scorch; dead leaves retained through winter.

Various fungal leaf and shoot diseases of *Prunus* species.

Apiognomonia erythrostoma causes leaves to develop brown blotches bounded by a yellow margin. Leaves will die but hang on the tree until the following spring, when new leaves will develop. The shoots and twigs remain healthy.

Distribution: Widespread and common.

Species affected: Prunus species.

Treatment: No control is necessary.

Blumeriella jaapii causes numerous small purple spots to appear on the leaves. Sometimes the rest of an affected leaf turns yellowish or orange-red but small green areas may persist around the spots. Some of the spots eventually fall out, leaving shot holes. Early defoliation may occur. In warm, damp conditions the fungus can appear as whitish felt patches on the undersides of the spots.

Distribution: Widespread and common.

Species affected: Prunus species.

Treatment: No control is necessary.

Monilinia laxa causes a disease known as blossom wilt, spur blight or wither tip, all of which accurately describe this disease. The fungus develops in the new shoots at the tips of the flowering spurs as they flush in the spring, killing the leaves and blossom. This can happen on isolated parts of the tree or over the whole tree in severe cases. It can appear similar to early frost damage, which also results in dead leaves and shoots hanging on the tree, although the distribution of the damage will not be the same. One difference is that the bark killing caused by this disease can extend a short distance into the previous year's shoot. Also, the affected blossom and fruits have a distinctive sweet smell of decay. Whilst the disease can completely ruin the blossom and fruit, long-term effects are limited. May affect other *Rosaceae* species.

Distribution: Widespread and common.

Species affected: Prunus species.

Treatment: Fungicidal sprays for *Molinia* will prove beneficial if the level of infection becomes a concern. See Table 1b.

Blumeriella jaapii.

Monilinia laxa

Apiognomonia veneta

Anthracnose of London plane

A fungal disease affecting leaves, shoots and twigs.

Several effects may be visible on trees. Spring shoots may suddenly die across the whole tree as if frosted, and individual twigs or branches may remain leafless or hold withered leaves. In summer, leaves can show extensive brown patches, mainly bordering the veins. Other leaves are affected mainly on their petioles, so that they are shed when still mostly green. Twigs may show small girdling cankers. It is likely that the fungus overwinters on the twigs and fallen leaves and affects the tree according to the weather patterns: a warm wet spring will result in poor flush; a colder spring may result in shoot death after flushing. Affected trees often refoliate throughout the summer and no long-term harm seems to occur.

Distribution: Widespread throughout the UK.

Species affected: *Platanus occidentalis* rather than *P. orientalis*, but the different forms of the hybrid London plane will vary in susceptibility.

Treatment: None generally necessary, but on important amenity trees fungicidal sprays can prove beneficial in controlling the effects of the disease. (See Table 1b)

Armillaria mellea/solidipes (=ostoyae)

Honey fungus

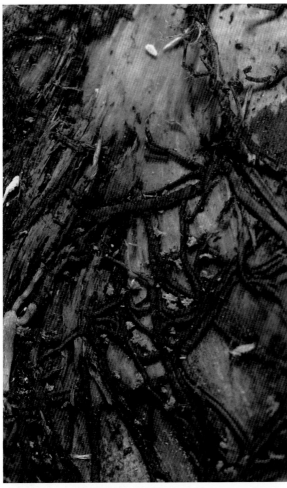

A fungus that can kill the cambial regions of the tree, also causing a white rot in the wood.

There are several species of honey fungus in the UK, only two of which are of arboricultural significance, *mellea* and *solidipes*. Honey fungus colonises through spore germination on wounds or by rhizomorphs (flattened root-like structures in the soil) that secrete enzymes, penetrating a patch of bark on a root and colonising and killing the underlying cambial zone and sapwood. Rhizomorphs extend through the soil from the original host and may travel many metres by utilising other woody nutrients in the ground to maintain energy levels. As a result, a number of trees and shrubs in a locality can be affected. Patches of exudates may appear on the lower trunk, and toadstools may appear at the base of the tree in many overlapping clumps. If a section of bark is removed at ground level a thick, felted white mycelium with a distinctive mushroomy smell may be found on the underlying wood. Toadstools have a honey brown cap, distinct gills and an obvious collar round the stalk below the cap.

Honey fungus kills a significant number of trees and shrubs, although it is more likely to affect trees already under stress, particularly drought stress. Due to the white rot element within the wood, investigations should be made as to structural integrity.

Distribution: Widespread throughout the whole of the UK.

Species affected: A wide range of broadleaf and coniferous species, with *mellea* more likely to affect broadleaf trees and *solidipes* conifers.

Treatment: None effective. Trees should be maintained in good condition by improving soil characteristics, mimicking natural woodland conditions wherever possible. Where infection has occurred resistant species should be planted. Removal of infected material may help but the wide spreading nature of the rhizomorphs makes this very hard to achieve fully.

Cameraria ohridella

Horse chestnut leaf miner

A leaf-mining moth whose larvae feed between the upper and lower surfaces (epidermises) of the leaf leaving damaged tracks or mines behind. The lower canopy is usually first to show symptoms.

Mines appear translucent. With repeated attacks mines can merge and reduce photosynthetic ability to the extent that the leaf dries, curls and falls prematurely. Early leaf fall can affect 70% or more of the tree which can appear autumnal by August. From a distance damage may be confused with *Guignardia aesculi* but on closer inspection is considerably different.

Adult moths appear from April onwards from overwintered pupae. They are only 5mm long and lay eggs along the lateral veins on the upper surface of the leaf. Larvae moult up to 5 times whilst feeding in the leaf over a 4-week period, and finally pupate for 2 weeks – although pupation can take up to 6–7 months when overwintering. Further generations are produced throughout the year: up to 5 have been recorded. Pupae overwinter in fallen leaf litter and are extremely tolerant of low temperatures.

Trees can become very disfigured quickly and repeatedly, leading to pressure for removal. However, there is no evidence to suggest a significant impact on tree health as they re-flush normally despite many successive years of leaf-miner damage. The moth will colonise an area quickly and can be transported many miles to set up new areas of infection by passive means: vehicles, trains and prevailing wind direction.

Distribution: Widespread throughout the UK and will spread to all areas.

Species affected: Principal host is *A. hippocastanum*. Other Asian horse chestnuts appear to be moderately resistant as do the majority of the North American species. Hybrids vary in resistance but *A x carnea* shows a significant resistance to this pest.

Treatment: See Table 1a.

Canker (general)

Fungal or bacterial infections of areas of the inner bark and cambial zone which can be initiated either at wounds or via other infection points such as leaf scars.

Cankers typically develop at certain times of year: in some cases in the growing season and in other cases in the dormant season, when host resistance is low. Some kinds of canker develop for only one year and can then become surrounded by a roll-like growth of new bark and wood. The resulting swollen appearance means cankers on twigs and small branches can easily be seen. Other kinds of canker develop year after year, killing successive rolls of new bark. Such cankers often have a target-like appearance because of the successive rolls of bark that have been killed. Occasionally, affected branches snap at points where cankers are present, especially if the canker encompasses a substantial proportion of the circumference. Affected branches or stems can also die as a result of being girdled by cankers. Partial girdling can result in poor condition with reduced vitality and leaf cover.

Bacterial canker of cherry, showing gum exudation and 'shot hole' effect on leaves.

Certain cankers, especially those caused by bacteria, involve not only the killing of bark and cambial tissue but also result in an excessive, tumour-like growth of bark tissue, which has a very rough, corky appearance. This occurs, for example, in the bacterial canker (aka 'tumour' or 'knot' of ash) caused by *Pseudomonas savastanoi* pv. *fraxini* and bacterial canker of poplar, caused by *Xanthomonas populi*. There is no such tumour-like growth in bacterial canker of cherry, caused by *Pseudomonas syringae* pv. *morsprunorum*; however, this bacterium not only kills bark but also causes a leaf spot, creating a 'shot hole' effect. It can spread from the infected leaves into the twigs via the leaf scars in autumn and via bark wounds. Gums exude from the damaged bark in early summer and the underlying wood is discoloured and non-functional.

Distribution: Widespread and common.

Species affected: A wide variety of trees affected.

Treatment: Pruning in midsummer may help reduce infection. Pruning tools should be cleaned after use to prevent cross-infection. Significant cankers can lead to increased incidence of branch failure and individual risk assessments should be made on trees in high value target areas.

Bacterial canker of ash.

Bacterial canker of poplar.

Chalara fraxinea

Ash dieback

A recently well-publicised fungal disease now established in the UK. Widely established in Europe, particularly Scandinavia and the Baltic countries, where it has caused widespread death of ash.

Various species of ash, including *Fraxinus excelsior* and *F. angustifolia*, are susceptible. Young trees generally die within a year or two after the first signs of the disease but large trees often survive for several years. In many areas, a small proportion of trees have remained free from the disease or have been only mildly affected and are thought to have some resistance. It is too soon to speculate on the effect of *Chalara fraxinea* on the UK ash population.

The disease starts in the leaves and spreads to the current year's shoots and then to older twigs and branches in later years. The initially affected leaves can remain attached for some time after withering. When the disease spreads into older twigs and branches, it forms cankers, which have a darkened, sunken appearance. A greyish-brown discolouration extends into the sapwood, which can be seen if the bark is removed from a canker. Heavily affected trees will have extensive shoot, twig and branch dieback. The fungus overwinters mainly on the midribs (rachides) and petioles of the fallen leaves, because the leaflets of ash usually disintegrate soon after leaf fall. If, however, the leaflets persist, their midribs can also harbour the fungus. During July to September (and sometimes into October) of the following year, pinhead-like fruit bodies of the fungus can form on the fallen leaf material in moist conditions (known at this stage as *Hymenoscyphus pseudoalbidus*), releasing spores that can infect the current year's leaves. Also, in very moist sites, there is evidence that the bark of trees can be directly infected via lenticels. The spores are thought to have only a short period of viability after being released but there is evidence that they can occasionally remain active long enough to cause infection of trees a considerable distance from the source. The average rate of spread on the Continent has been 20km to 30km per year but this has been accelerated by the movement of infected plants.

Once affected, trees cannot be cured and their condition should be reported and monitored for continued health.

Distribution: Now established in the UK. Widely established in Europe, particularly Scandinavia and the Baltic countries.

Species affected: Various species of ash, including *Fraxinus excelsior* and *F. angustifolia*, are susceptible.

Treatment: Chalara is being treated as a quarantine pest under national emergency measures and suspected cases should be reported to the relevant authorities (see page 79). At present all imports of ash seeds and plants into the UK are prohibited as well as all movements of the same throughout the UK, whether grown here or not. Ash wood and logs may continue to be moved, except from sites where a statutory Plant Health Notice has been served. Timber may be moved from these sites with the appropriate authority.

Cinara cupressi

Cypress aphid

A sap-sucking insect that is active from May to November.

Symptoms include rapid yellowing and finally browning of affected parts. By the time browning occurs the majority of the aphids will have disappeared. Effects are most obvious on clipped hedges. Checks should be made through the spring and summer the following year to confirm the presence of the aphids. These are generally large, distinctive aphids up to 4mm long with two dark stripes on their backs.

These aphids reproduce through live birth during the summer and whilst the majority of overwintering seems to occur in egg form it is clear that some adults also overwinter.

Distribution: Common throughout southern England and reported in Scotland.

Species affected: Common on Leyland cypress, particularly 'Castlewellan Gold', as well as Lawson cypress, Monterey cypress and some *Thuja*.

Treatment: Spraying with an appropriate insecticide to coincide with adult feeding can control this pest. May only be practical on clipped hedges. See Table 1a.

Cossus cossus

Goat moth

A large evening-/night-flying moth up to 70–80mm in wingspan.

The goat moth lays its eggs in the bark of trees in early summer. These hatch and the young larvae feed in galleries under the bark during the autumn and winter. The following spring the adult caterpillars bore into the wood creating large tunnel networks and spend the next year, sometimes two, maturing within the tunnels. Pupation takes place in these tunnels the following spring, surrounded by frass. Pupation lasts for a month before the mature adult moth emerges in the early summer.

Distribution: A widespread but uncommon pest.

Species affected: Many broadleaved trees, particularly fruit trees.

Treatment: See Table 1a.

Cryptococcus fagisuga

Felted beech scale

A sap-sucking insect that is attached to the tree by its mouth parts and moves only in its first instar.

The adult lays eggs in mid to late summer that hatch into 'crawlers' which generally disperse a short distance and start to feed and become stationary soon after.

Significant infestations can look unsightly, can reduce the vitality of individual trees and may make the tree susceptible to beech bark disease (*Neonectria coccinea*) as a result. There seems to be some variation in susceptibility with stands of beech showing only selected individuals affected, but such differences are partly due to different rates of build-up of the infestation, which can takes many years.

Distribution: Widespread and common.

Species affected: Beech.

Treatment: Physical removal by scrubbing with a mild detergent can control numbers.

Cryptostroma corticale

Sooty bark disease

An endophytic, fungus that can apparently exist in a latent state in the wood of healthy sycamores for many years.

The fungus is triggered into activity within the sapwood by moisture stress associated with long, hot summers. The active spread of the fungus in the sapwood can be extensive enough to kill the entire tree.

The tree may fail to flush or leaves suddenly die but they are held on the tree in the summer as dried, up-turned specimens. Branches cut across their section reveal a green-brown blotchy stain often with a black margin, usually only visible whilst the disease is active. The fungus spreads from the sapwood into the overlying, now dead, bark, the outer layer of which becomes raised and then peels off, revealing a black sooty layer of spores. Owing to the latent nature of the fungus, it might be present in a substantial part of the sycamore population. However, outbreaks are sporadic and tend to have only a local affect.

Distribution: More common in the warmer southern areas but distribution may change with climate change.

Species affected: Almost exclusive to sycamore.

Treatment: None available, but avoid damaging the rooting area, thus causing water stress.

Dendrolimus pini

Pine tree lappet moth

A large moth endemic in Europe and Central Asia with the potential to have a devastating impact if it becomes established in the UK.

Occasional sightings have been made in southern England where it is thought to have been brought in on imported trees or swept over from the Continent. During the middle of the last decade a small population may have established near Inverness and was subject to eradication measures. This work is ongoing.

The moths emerge in late June to mid-August and mate on the lower trunk. The females climb the tree to lay eggs on needles, twigs or the bark of Scots pine. Eggs hatch during August and September and feeding commences on the edges of needles. After the first frost the caterpillars climb down to overwinter under the leaf litter close to the tree. As soil temperatures increase the caterpillars return to the canopy and resume feeding. This time feeding is voracious and whole needles, particularly old needles, are fully consumed. Caterpillars can reach 80mm long. Pupation lasts for 4–5 weeks in the early summer in semi-transparent cocoons in the canopy, on twigs or on understorey vegetation.

Distribution: See entry above.

Species affected: Mostly Scots pine, but other pine may be affected and in preferential conditions spruce and Douglas fir may also be affected.

Treatment: None available at present.

Note: Any suspected outbreak of pine lappet moth should be reported to the relevant authorities (see page 79). If possible trap and retain the specimen for examination.

Dothistroma septosporum

Red band needle blight

A fungal disease affecting a range of conifers, with pines – particularly Corsican and lodgepole – being very frequent hosts.

Infected needles typically develop yellow and tan spots and bands, which soon turn red. The ends of the needles then turn reddish-brown whilst the needle base remains green. Needle symptoms are most apparent in June and July when the spore-containing fruit bodies are most abundant, after which the infected needles are lost and trees can have a typical 'lion's tail' appearance, with only a tuft of the current year's needles remaining at the branch ends. Defoliation can occur year on year, reducing vitality, and may therefore allow other pathogens to affect infected trees.

Distribution: Widespread and common.

Species affected: A range of conifers, with pines, particularly Corsican and lodgepole, being common hosts.

Treatment: None available.

Drepanopeziza sphaeroides

Anthracnose of willow

A fungal disease of willow that can be severely disfiguring. Only affects weeping willow forms and some ornamental contorted varieties.

The disease occurs on shoot and leaves, killing and blackening the tissue and causing leaves to drop in severe cases. As the shoot hardens, small persistent cankers may appear. The fungus will overwinter on fallen leaf litter and within the cankers ready to reinfect next season. Severe attacks can reduce the growth of specimen trees, as well as disfiguring them.

Distribution: Widespread and common.

Species affected: All weeping and contorted varieties of willow may be affected.

Treatment: Subject to current regulations, appropriate fungicides can control the disease on particularly important trees. Removal of leaf litter and fallen twigs will reduce the ability to reinfect. Alternatively, replace with more resistant species. See Table 1b.

Erwinia amylovora

Fireblight

A bacterial disease on a wide range of trees and shrubs in the *Pomoideae* section of *Rosaceae* (e.g. apple, pear, whitebeam and hawthorn), killing shoots and twigs and potentially whole trees.

The bacteria overwinter around the margins of small cankers, forming a gummy ooze in which they are distributed by rain, pruning tools, birds or pollinating insects. Infection occurs via the flowers in spring and leads to rapid spread throughout the phloem and cambial zone of the affected twigs and branches, which soon die as a result. This happens typically after periods of warm, wet, windy weather around flowering time. The leaves of the affected branches remain attached in a blackened state, similar to the effect of fire damage. The distribution of such damage in the crown of the tree is, however, more sporadic than in the case of fire damage and the diagnosis of fireblight can be confirmed by the presence of dead, reddish-brown inner bark, revealed by paring away the outer bark just beyond the attachment points of the dead twigs and branches.

Distribution: Throughout the southern counties and Midlands of England, and Wales. Also established in central Scotland. Sporadic outbreaks occur in other areas. Ireland, Isle of Man and the Channel Islands do not have the disease.

Species affected: Trees and shrubs in the *Pomoideae* section of *Rosaceae*, e.g. apple, pear, whitebeam and hawthorn.

Treatment: Cutting out the infected material early in infection to a point well beyond any staining can effectively control spread of the bacteria within the tree. Note: cutting equipment must be sterilised between each cut. Burn infected arisings. Late autumn/winter washes with a liquid copper solution will help if infection levels are high. See Table 1b.

Note: In areas known for outbreaks, ensure that planting schemes are not dependent upon susceptible species.

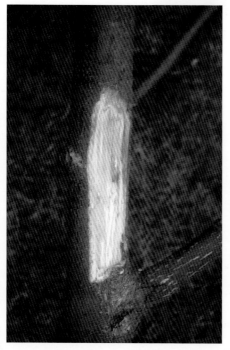

Erysiphe alphitoides *(also known as Microsphaera alphitoides)*

Oak leaf powdery mildew

A fungal disease affecting leaves, particularly the second, lammas, flush.

Particularly common in mid to late summer as the second flush of leaves, the lammas growth emerges, a greyish white fungal bloom may be seen covering the leaves. Erysiphe can attack all young leaves and soft shoots, covering them with a felty-white mycelium, causing them to shrivel and blacken. Mild overcast conditions are optimal for development of the disease, which is usually most prominent in summer when warmer conditions prevail.

Distribution: Widespread throughout the UK.

Species affected: All deciduous oak can be affected.

Treatment: None generally necessary, but on important amenity trees fungicidal sprays can prove beneficial in controlling the effects of the disease. (See Table 1b)

Euproctis chrysorrhoea

Brown tail moth

A moth whose caterpillar can defoliate a wide range of trees and shrubs.

Adult moths mate in midsummer, laying eggs on the underside of leaves, mostly hawthorn or related species. The eggs hatch later in the summer when the caterpillars emerge and feed and moult. These caterpillars are about 30mm long, have 2 red spots on their backs and are covered in tufts of brown hair. These hairs break off easily and are an irritant to humans, causing asthma-like symptoms and eye irritation. The caterpillars overwinter in 'tents' of web-like material and emerge in the spring to continue feeding. In late June the caterpillars pupate to emerge in July and start the process again.

Distribution: All counties of England are affected but particularly southern coastal regions.

Species affected: Mostly hedgerow tree and shrub species.

Treatment: During winter the 'tents' can be pruned out and destroyed. Care should be taken to use suitable PPE to avoid contact with the hairs. Appropriate insecticide treatment during feeding periods can reduce caterpillar numbers. See Table 1a.

Guignardia aesculi

Horse chestnut leaf blotch

A fungal disease killing parts of the leaves, most commonly affecting the leaf margins.

The symptoms may be obscured due to the increase in the frequency and severity of attacks by horse chestnut leaf miner.

Typically the leaves look tattered and as if early autumn has set in by late summer. Severely affected leaves will curl along their length and premature leaf fall is common. The fungus appears to reinfect trees in the spring from fallen leaves. Whilst it may appear dramatic, there is no evidence of long-term harm.

Distribution: Widespread and common.

Species affected: Horse chestnut and red chestnut are commonly affected in the UK with reports of other *Aesculus* species being affected in other countries.

Treatment: None required although with isolated trees the removal and destruction of fallen leaves may reduce the impact of the disease the following year.

Leaf spots and blotches (general)

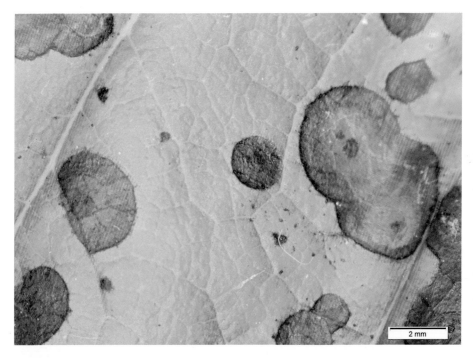

Various fungal or other infections of leaves will result in blotches and spots and may result in premature leaf fall.

Leaves may become discoloured, withered, distorted and may be retained as shrivelled dead leaves, or be shed when still apparently green and healthy.

Such damage may affect localised parts of the crown or the whole crown. Some of the diseases concerned usually begin in the lower crown and work upwards, partly because they are caused by fungi that release most of their spores near ground level. Small fungal structures may be visible with the aid of a hand lens on the affected leaves, often near the veins and midribs, but these might develop only in the following spring, depending on the fungus concerned. Many of these diseases tend to be most prevalent after periods of warm, wet weather.

Repeated attacks of a leaf disease can weaken the resistance of a tree to other diseases, some of which can affect the twigs, branches or the whole tree. If a leaf disease starts early in the season, a second flush of foliage may occur in response. This may fail to 'harden off' and therefore be vulnerable to winter cold or to other diseases. In the absence of such secondary effects, leaf diseases on deciduous trees are not usually a threat to the survival of the host. They can, however, spoil the appearance of amenity trees. Evergreen trees or shrubs (especially conifers) can be more severely affected since they generally store a higher proportion of their food reserves in their foliage.

Treatment: Many leaf diseases are caused by fungi that overwinter in fallen leaf litter. Removal of this litter may help reduce the severity of attacks in following years but it is usually less effective in the case of fungi that can additionally overwinter elsewhere (e.g. in bud scales on the tree). The application of appropriate fungicides may help, as may improving air circulation through a plant by selective pruning. See Table 1b.

Molinia mespili on medlar.

Venturia inaequalis on crab apple.

Lymantria dispar dispar

European gypsy moth

A moth whose caterpillars feed on a wide range of broadleaved trees and can cause severe defoliation.

The female moth cannot fly and so lures mates – the males can fly freely – and lays eggs in a location close to emergence. Egg masses may be found on trunks, main branches, cordwood etc. in late summer and the larvae develop and overwinter inside the egg cases.

Eggs are buff coloured and covered in hairs from the adult's body to help insulate them over winter. Caterpillars hatch in spring and climb to the leaves to feed and disperse. This is achieved by the spinning of a fine silk thread to catch the wind. Caterpillars may float up to a mile in this fashion. Once dispersal is complete, maturation feeding and moulting occur. The caterpillar will typically produce 4-5 instar phases, enlarging each time until 50-90mm long, covered in hairs and with blue and red dots on the head and tail respectively. The caterpillars continue to feed on the leaves of a range of broadleaved trees, notably oak. The caterpillars pupate on trunks, main branches etc. in the ridges of the bark in mid-June to emerge 2-3 three weeks later as fully formed adults.

Distribution: Uncommon if not rare; may only be established in parts of East Anglia and Greater London.

Species affected: Many broadleaf tree species.

Treatment: See Table 1a.

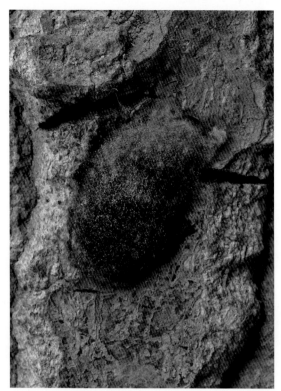
Egg mass, approximately 1-2 cm in diamter.

Melampsora or *Melampsoridum spp.*

Rusts of poplar, birch and willow

Rust on poplar, showing upper and undersides of leaves.

These rusts mainly affect the leaves, but in certain varieties of willow infection can spread to twigs.

Infection occurs usually in late summer but sometimes starts in early summer, especially where sources of infection are nearby.

The affected leaves show yellow or orange pustules, most often on the undersides. There is usually a yellowish discolouration of the tissue on the corresponding upper leaf surface. The pustules release vast numbers of airborne spores which cause several cycles of infection in a single season. The fungus does not directly kill the leaf tissue but the affected leaves often turn generally yellow and are shed prematurely.

By the time of shedding, the leaves bear small, shiny, blackish overwintering structures, from which spores of a different type are released in the following spring. These spores can infect only an 'alternate host' (usually a conifer or an herbaceous plant) which is host-specific to the particular species of rust fungus concerned. The infected alternate host produces spores of yet another kind that reinfect the original host tree later in the year. Some of the rust fungi involved may directly reinfect the primary host from overwintering infection of buds or twigs, mainly in birch and willow.

Heavily infected trees show a reduced growth rate and can become susceptible to other diseases, owing to a lack of hardening in the current year's growth. Poor hardening can also render products such as withy rods useless. Very heavy infection of poplars and willows (as can occur when they are grown at high density) can lead to a severe reduction in growth or death.

Distribution: Widespread throughout the UK.

Treatment: Severe economic loss could be reduced by the use of fungicides, subject to current regulations. See Table 1b.

Melampsoridium betulinum – rust on birch.

Rust on poplar.

Neonectria coccinea

Beech bark disease

A weakly pathogenic fungus that can kill small patches of bark or occasionally much larger strips.

In the latter case, or if the small areas of bark death coalesce into larger areas, the overall health of the tree can be affected, sometimes culminating in death. Also, the underlying wood is exposed to colonisation by beetles and/or other fungi.

The fungus is commonly present in woodland and causes damage only when the tree has been physiologically stressed by heavy infestation with the felted beech coccus (see *Cryptococcus fagisuga*) or by severe drought.

During the early stages, small black weeping spots appear on the main stem. The exuding liquid usually dries out to form small tarry encrustations. These spots represent the death of patches of the underlying inner bark, which can be revealed as an orange-brown discolouration with a characteristic smell if the outer bark is pared away. Bright red pinhead-sized fruit bodies of the fungus may appear on the surface of the dead bark. If the dead patches remain small and localised, the tree usually survives but develops a rough, cankered appearance where new bark and wood have developed around the dead patches (see Cankers). If extensive areas of bark are killed, the crown of the tree can become sparse. Severely affected trees can die, shedding large patches of bark in the final stages.

Distribution: Widespread and common.

Species affected: Beech.

Treatment: In cases where the latter stages are reached, there is no effective control therapy. The removal of the beech coccus by physical scrubbing may reduce the likelihood of initial disease development and will also reduce the direct adverse effects of the coccus infestation. Trees may recover from mild infections if growing conditions are favourable.

Close up view of clusters of minute fruiting bodies.

Oak decline

Acute oak decline (AOD)

Agrilus biguttatus – associated with AOD.

A new, little understood decline of oak resulting in stem bleeding and dieback which is thought to be associated with certain bacteria.

It affects native oaks generally over 50 years old. The key symptoms are black, weeping patches on stems (bleeds), which result from the death of the underlying tissue. Larval galleries of the buprestid beetle, *Agrilus biguttatus*, are usually associated with the bleeds and can often be seen weaving a sinuous path in the cambial zone when the bark is removed. The beetle and the disease often occur together on the same tree, but the disease can be identified where the death of the tissues extends well beyond the galleries. Unlike trees showing chronic oak decline, some of those affected with AOD die within 4 to 5 years of the onset of symptoms. In the early stages of this condition no changes in canopy health are noticeable but as trees approach death the crown may be visibly thinner.

Chronic oak decline

A complex mix of causes, both biotic and abiotic, that slowly results in a decline in health of trees over many years.

Prolonged droughts, repeated episodes of defoliating insects and leaf colonisation by powdery mildew, as well as fungal root colonisation (e.g. *Collybia*, *Armillaria* and *Phytophthora*), can all contribute to this long-term decline. Trees may undergo retrenchment, which leaves a smaller, lower crown with deadwood protruding from the upper crown (stag-headed).

Distribution: Mostly in the Midlands and south-east but spread is being reported from many other areas.

Species affected: English oak.

Treatment: None practical at present; keeping soil conditions optimal, to include mulching, may help delay the onset of these conditions.

Ophiostoma novo-ulmi

Dutch elm disease

A fungal wilt disease that disrupts the water transport system of a tree.

Spread by elm bark beetles or through root grafts or common rootstocks.

The beetles colonise the bark of dying or very recently dead elms and the fungus grows within the breeding galleries. The new generation of emerging beetles take spores with them to infect healthy trees at feeding points high in the crown. Infection leads to the blockage of xylem (the water transport vessels), resulting in wilting of the tips of branches and then entire branches throughout the crown with associated yellowing then shrivelling of leaves. This normally starts in the upper crown and extends to the lower branches.

Infection through grafts is normally very rapid with symptoms developing rapidly over the whole crown, resulting in tree death.

Distribution: Widespread throughout the whole of the UK.

Species affected: Elm and *Zelkova* spp. although the latter are considered more resistant to the disease. Various breeding programmes have released resistant species of elm but their long-term performance in the UK is unknown at this time.

Treatment: Prompt removal of infected trees to prevent the beetles breeding is an effective form of control.

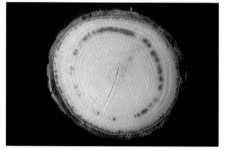

Phytophthora species

Phythophthoras are fungus-like organisms most closely related to brown algae. All are transmitted in a waterborne environment, be it groundwater, open water, rain splash, mists or spray drift.

Aerial and root-infecting species are recognised as separate entities. The latter generally colonise the roots and lower stems of many shrubs with a wide range of tree species being commonly affected. External symptoms often include bleeding patches on the lower stem in the form of dark coloured liquid weeping from the bark. This exudate dries to a brittle crust. When bark is pared away, the underlying inner bark (phloem) tends to be mottled and discoloured, ranging from a pinkish hue through to orange or brown-black, depending on the affected plant species. The wood may also be stained a blue/black colour. Typically, tongues of infected bark extend up from affected roots or root collar. Loss of roots and partial girdling of the root collar also lead to crown symptoms and dieback.

Examples of root-attacking Phytophthora include:

- *P. alni* – affects alder species.
- *P. cactorum* and *P. citricola* (now called *P. plurivora*) – often the cause of basal bleeding patches on species including lime and beech.
- *P. cinnamomi* – affects species including sweet chestnut, beech and yew.
- *P. lateralis* – affects Lawson cypress and associated species and cultivars.

Aerial Phytophthoras attack the above-ground plant parts such as leaves and shoots, and some species can infect through intact mature bark. When bark is infected the symptoms are similar to those of root-attacking Phytophthoras, except that the bleeding patches tend to be higher up the trunk and on branches. Affected foliage and shoots can have blackened spots and lesions, and typically spores are produced abundantly from this infected green tissue.

Examples of aerial Phytophthoras include:

- *P. ilicis* – affects holly causing premature leaf loss.
- *P. ramorum* (known as sudden oak death in the USA, where it affects black oaks and tanoaks). In the UK the principal shrub host is rhododendron. Beech, sweet chestnut, turkey oak and red oak have all been confirmed as hosts but not on a wide scale. The most significant host in the UK is Japanese larch, a major timber tree, with confirmed infection along much of the west coast. Larch and rhododendron are the highly significant hosts because their infected foliage produces a huge number of spores, causing disease spread. Most other tree species produce far fewer spores when infected, so pose little risk.
- *P. kernoviae* – a species of *Phytophthora* distinct from *ramorum* but also associated with rhododendron and beech. Although considered to be more pathogenic than *P. ramorum*, it is almost exclusive to Devon and Cornwall.

Note: *P. ramorum* and *P. kernoviae* are listed quarantine organisms. Their suspected or confirmed presence must be reported to the relevant authorities (see page 79).

Treatment: Currently there is no full understanding of all the Phytophthoras present in Britain. Some trees recover from infection, particularly with root-infecting species, others can die rapidly. As the majority of transmission, particularly for root infection, is through groundwater, site drainage may play an important part in limiting disease development and preventing spread.

Pseudomonas syringae pv. *aesculi*

Bleeding canker of horse chestnut

A bacterial canker that causes bleeding from the bark of any part of the trunk or main branches in trees of all ages.

The common horse chestnut and red horse chestnut are particularly susceptible with other species being less so, notably *Aesculus indica* and *Aesculus flava*. Young trees are particularly vulnerable as necrosis can completely girdle the main stem in a short time. Mature trees can be seriously disfigured, owing to the death of large sections of the crown as a result of disease in the supporting vascular tissues. Even very large trees can be killed in a few years but there are many trees in which the disease has stabilised, remaining only in localised patches of bark. Individual trees vary in susceptibility, some succumbing very quickly whilst others either never develop the symptoms or recover after an episode of the disease.

In the early stage of bark-killing, the only externally visible sign is usually the appearance of scattered bleeding points, from which drops of gummy liquid ooze. When this exudation begins, often in the spring, the liquid is dark but transparent, but later it becomes more opaque and acquires a rusty red, yellow-brown or blackish colour. As the fluid accumulates during the summer it can run some way down the tree before drying to leave a dark, brittle crust near the point of exit. Renewed bleeding may be seen later in the year, often in autumn. This suggests that pathogen activity is greatest under the moist, mild conditions of spring and autumn.

In some cases, the bark death associated with bleeding extends for many metres along the length of the affected branch or stem. The branch or stem may eventually die as a result of continued bark-killing, where this leads to girdling. The affected areas of inner bark can be seen by paring away the outer layers. They show an orange-brown discolouration, containing numerous dark lines, where the host appears to have laid down temporary barriers to the spread of the bacteria. A spreading lesion will usually show a diffuse edge, while a stabilised lesion will have a sharper boundary with the surrounding healthy bark.

The extent of the lesions becomes increasingly apparent after several months, when the affected areas of bark begin to crack and eventually fall away. Strongly developing rolls of new bark and wood develop along the edges of the exposed wood if the disease is stabilising. Secondary decay can set in later, as shown by the appearance of fruit bodies of wood-rotting fungi.

Distribution: Widespread and common.

Species affected: Common horse chestnut and red horse chestnut are particularly susceptible with other species being less so, notably *Aesculus indica* and *Aesculus flava*.

Treatment: There is currently no commercially available treatment. Affected trees should be individually monitored in order to see whether their condition stabilises. The risk of spreading the bacterium between trees should be reduced by sterilising tools and by burning bark and arisings from affected trees. Improving soil conditions and thereby the vitality of the tree can help the recovery of infected trees.

Pulvinaria regalis

Horse chestnut scale

An insect pest of horse chestnut, maple, lime and many other broadleaved trees.

This sap-sucking insect appears as a waxy white circular scale with a brown 'head' on the branches and stems of affected trees. These are generally the dead female adults that have laid their eggs and provide a protected environment for the young that will hatch, move onto the leaves to feed and then mature over winter on the twigs before moving back onto the main limbs and trunks to lay the next generation of eggs. These insects, whilst unsightly, cause little harm to an established, well-growing tree, but they will reduce vitality. In an already stressed tree scale colonisation can contribute further to decline.

Distribution: Common in all urban areas in England.

Species affected: Horse chestnut, maple, lime and many other broadleaved trees.

Treatment: Treatment is difficult because the timing of applications needs to coincide with hatching, although a more systemic application may be possible. Physical removal of the dead females is unlikely to control the insect but may improve appearance.

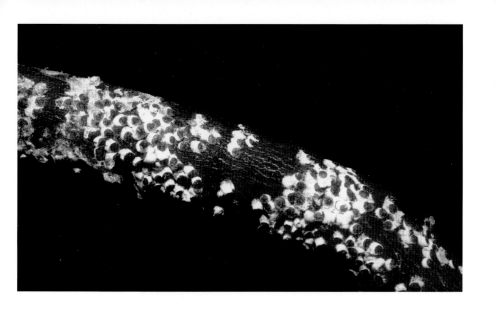

Female scales are approximately 5-7mm long.

Rhytisma acerinum

Tar spot of sycamore

A fungal disease of sycamore leaves.

The fungus causes obvious black, circular spots on the leaf surface. A common but unimportant disease of sycamore with most trees in the UK affected. Reinfection occurs from the previous year's leaf litter.

Distribution: Widespread throughout the UK.

Species affected: Almost exclusive to sycamore but other maples may be affected.

Treatment: None necessary, but clearing the previous year's leaves may delay or reduce onset.

Cristulariella depraedans

A fungal disease of sycamore leaves.

An obvious leaf disease of sycamore causing small white spots over the whole surface area.

Leaves may fall prematurely. Striking but of little arboricultural significance.

Distribution: Widespread.

Species affected: Mostly sycamore but records exist on other maples.

Treatment: None required.

Tar spot of sycamore.

Seridium cardinale

Coryneum canker

A fungal canker that girdles and kills small branches and can spread to significant sections of a canopy. Primarily found on Monterey cypress but other cypresses may be affected.

Yellow or brown patches of foliage develop in the crown as girdling increases. Entire branches can lose foliage, so that sections of the crown appear brown. The dead sections are especially conspicuous in Monterey cypress owing to the tendency for branches to be isolated within the crown. The death of bark on the affected branch is often accompanied by resin bleeding from the surrounding tissues. The disease spreads slowly within the tree but can ultimately kill it.

The fungal spores are spread in water or by physical movement (birds, insects) and infect through natural fissures in the bark, e.g. branch crotches. The fungus then develops, girdling branches and spreading into the stem, which can eventually become girdled.

Distribution: Widespread and common.

Species affected: Mostly Monterey cypress but other cypresses may be affected.

Treatment: Cutting out the affected areas, well beyond the apparent limits of infection, may slow the spread of the disease on trees showing recent symptoms. On trees which have been showing symptoms for some time the extent of pruning required is likely to be too great and the likelihood of missing affected portions too high for this to be effective.

Sesia apiformis

Hornet moth

A wasp-like moth that emerges in midsummer from the base of poplars, leaving a ragged 8mm exit hole.

Eggs are laid at the base of poplars with the hatching larvae burrowing under the surface of the bark and early wood and spending up to 2 years in maturation feeding before emergence.

The moth itself is considered a secondary agent of decline, but in certain areas there is a concern that it is acting as a primary pest and killing otherwise healthy trees. In any event, repeated attacks may weaken the vitality of the tree and allow other pathogens to gain entry.

Distribution: Confined to the southern half of Britain below the Midlands at present.

Species affected: Poplar.

Treatment: See Table 1a.

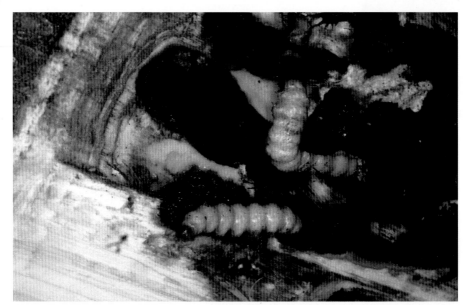
Hornet moth larvae.   Louis-Michel Nageleisen, Département de la Santé des Forêts, Bugwood.org

Hornet moth damage.   Gyorgy Csoka, Hungary Forest Research Institute, Bugwood.org

Splanchnonema platani

Massaria

A fungus that is thought to be host-specific in London plane and that was previously thought to be mainly endophytic or weakly parasitic on twigs and small branches.

More recently it seems to affect small branches, typically killing a strip of bark and sapwood on the upper side of the branch. Subsequent decay of the affected sapwood, perhaps by other fungi, often leads to branch failure. Drought conditions are thought to make the tree more susceptible and, if unchecked, can lead to the infection of larger branches.

Since the disease is initially confined to a strip running along the upper quadrant of a branch, often the affected strip cannot be seen from ground level. Typically, the strip is broadest near the base of the branch and tapers to a point. The disease is most frequently found on shaded branches in the lower-crown and mid-crown regions. Relatively small diameter branches, up to 150mm in diameter, can be killed by the disease in one growing season. Branches will retain dead, withered leaves with bark peeling on the upper surface. Larger diameter branches may not exhibit these symptoms. In either case, branch shedding can occur quite soon after the onset of the disease.

Distribution: Mainly restricted to London.

Species affected: *Platanus* species, cultivars and hybrids.

Treatment: For trees in high target areas, climbing inspections may be appropriate to identify the presence of Massaria canker. Where it is confirmed, proportionate tree management – both physical and moisture related – may be required to reduce stress on the affected parts of the tree. Improved drainage and appropriate mulching to allow groundwater availability may help reduce the impact of this disease.

Taphrina spp.

Witches' broom

Witches' broom on birch - *Taphrina betulina*.

A fungal disease of little significance causing abnormal twig proliferation and reduced flowering in the affected parts.

Takes the form of a tight cluster of shoots orginating from a common location on a branch or stem, often forming a dense sphere or clumb of twiggy growth that gives it the common name.

Distribution: Widespread and common.

Species affected: Common on birch, hornbeam, cherry and various conifers.

Treatment: May be pruned out if considered unsightly or has become so large that it is unstable.

Witches' broom on cherry – *T. cerasi*.

Witches' broom on hornbeam – *T. carpini*.

Thaumetopoea processionea

Oak processionary moth

A major defoliator of oak in Europe, it has now been established in London for several years.

Eggs are laid in late summer on twigs and small branches high in the canopy in 'plaques' consisting of up to 200 eggs covered in greyish scales; they overwinter in this form. Hatching occurs in April. Larvae feed on the foliage of the tree in groups and when not feeding, congregate in a communal nest made from white silk webbing. The larvae follow each other, nose to tail in long processionary lines between nest and feeding areas. Larvae moult 5 times, shedding their skin between each phase in their communal nest. Older caterpillars tend to spend the majority of their time in the nest and are therefore less controllable by chemical application. The caterpillars have distinctive long white hairs emerging from reddish 'warts' along their bodies, a single dark stripe down their backs and a whitish line along each side. The small brown hairs that cover the rest of their bodies cause respiratory problems and skin rashes in humans.

Caterpillars pupate after 9–12 weeks of feeding and remain in their nests for approximately 2 weeks before emerging in July or August to start the process again. Adult moths fly at night and are well camouflaged.

The hairs on the more mature caterpillars are irritating due to a chemical that is potentially harmful to humans. Allergic reactions are very common, causing rashes, conjunctivitis and respiratory problems. Health problems can occur even when there is no direct contact with the caterpillars as hairs break off easily. Abandoned nests should be treated with great care as they are full of hairs from shed skins etc.

Distribution: Mainly London.

Species affected: Oak.

Treatment: Cutting out, vacuuming or burning nests can be effective in reducing number as most caterpillars will return to nests during the day. Protective clothing is recommended. Chemical applications can be effective at the early larval stage when larvae feed continuously (see Table 1a). Further information specifically relating to control of this pest can be found on the Forestry Commission website.

Venturia spp.

Scab

Following cool, wet weather, black spots and blotches appear, caused by a fungal infection resulting in whole leaf death.

Leaves are retained on the tree, shrivelled but firmly attached. Lesions appear on the shoots resulting in blackening and curling. A velvety green/black mould appears on leaves and shoots. Scab can cause significant disfiguring of the tree and loss of vitality.

Distribution: Widespread and common.

Species affected: Common on *Malus*, *Sorbus* and *Populus*; to a lesser extent on *Pyrus* and *Salix*.

Treatment: Fungicidal sprays can provide effective control but need frequent application during the growing season. Resistant varieties are available where replacement is a reasonable alternative. See Table 1b.

Scab on willow showing dieback of shoot.

Zeuzera pyrina

Leopard moth

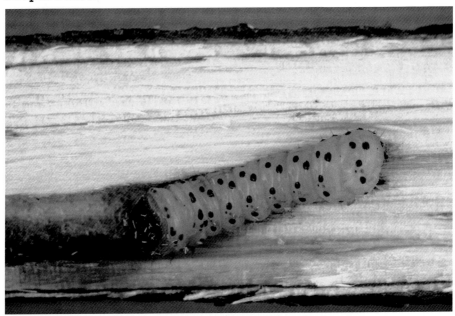

A large moth, generally white with blue spotting and up to 60mm in wingspan.

It flies at night but may be found resting during the day. The females lay eggs throughout the summer in bark cracks etc. The caterpillars disperse to feed on leaves, petioles and soft shoots. As they mature they move onto twigs and bore down centrally into small branches and finally larger diameter branches and stems. This can result in rapid wilting and fracture of affected parts. Feeding areas can change over the course of development until final pupation in a larger diameter stem. Maturation feeding can take 2–3 years until the caterpillars pupate in the spring. New adults emerge in midsummer to repeat the cycle. Adults live for 1–2 weeks.

Distribution: Southern England and southern Wales.

Species affected: Wide range of broadleaf trees and shrubs.

Treatment: Affected parts can be pruned out and destroyed. Appropriate pesticides may be employed: see Table 1a.

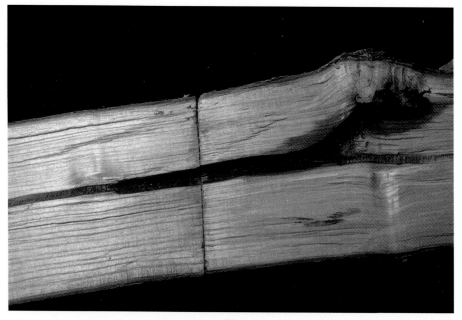

Suggested management programme for the control of pests and diseases

Control of pests and diseases can be demanding. There is no 'magic bullet'. Management strategies should be aimed at promoting tree vitality as well as treating the pest and/or disease in question. The following management strategy is recommended:

1. Inspect for any external symptoms that could induce stress in trees, e.g. new building construction, and remediate if necessary.
2. Ensure optimal tree nutrition. Sample soils for nutrient and pH levels. Based on a soil nutrient analysis, fertilise with the appropriate soil nutrients.
3. Apply a suitable plant protection product outlined in Table 1 to control the pest/disease in question if it is deemed that the extent of infection is sufficiently high to be detrimental to the tree.
4. Apply organic matter, such as undecomposed wood mulch, to a 5–10cm depth and not heaped against the base of the stem. Mulches should be applied to 1m beyond the canopy drip line if possible.
5. Guard against over- and under-irrigation. Use soil moisture levels to ensure soil water status is optimal for tree growth.
6. Decompact soil if bulk density values are $\geq 1.34 g/cm^3$ using an air-spade to facilitate root growth.

Research shows that adoption of this strategy provides a greater chance of tree recovery and survival from attack and infection compared to no management strategy.

Many of these measures will be new to professionals involved in the management of trees within towns, cities and parks. They will require changes to our existing management programmes and perhaps this is the key. If we don't adapt our current management systems to embrace these new technologies to counter existing as well as emerging pests and diseases, then many of our dominant UK landscape trees may go the way of the elm and disappear over the next 40–60 years.

The application of any pesticide must be informed by the manufacture's guidelines and the Control of Substances Hazardous to Health Regulations 2002. Pesticide must be applied by suitably qualified and competent personnel who hold the correct certification and have completed a full risk assessment.

Table 1a: Plant protection products registered for use in an amenity environment – Insecticides

Product	Active ingredient	Comments and pests controlled
Bandu	Deltamethrin	A broad-based synthetic pyrethroid insecticide which controls aphids, oak processionary moth, sawflies, scale insects, caterpillars and thrips. Apply at the first signs of damage with follow-up treatments at 6–8 week intervals if required. Spray is rainfast within 1 hour.
Dimilin Flo	Diflubenzuron	An insect growth regulator primarily for the control of moths (oak processionary, brown tail, bud, carnation, tortrix, codling etc.). Also controls rust mite in fruit trees and earwigs. Most effective control achieved by spraying from egg lay to early instar stage. Negligible effect on many beneficial insects (ladybirds, spiders, predatory mites).
Admire	Imidacloproid	A broad-based systemic insecticide registered for the control of chewing and sucking insect pests (aphids, horse chestnut leaf miner, flea beetles, sawflies, scale insects, caterpillars, thrips) of amenity trees. Can only be applied in the months February to March and October to November. Imidacloprid is slow to work but once established, 1–2 years' control can be expected.
Spray Oil		Primarily for the control of soft-bodied insects, e.g. scale, whitefly, mealy bugs, spider mites, aphids. Acts by physical and not chemical means. Can be used as an overwintering spray to control some insect eggs. Useful product against mite attack.
Savona	Fatty acids (soap)	Registered for the control of insects including scales, whitefly, mealy bugs, spider mites, aphids.
DiPel DF	*Bacillus thuringensis*	DiPel is comprised of live bacteria that produce a toxin which controls caterpillars when ingested by the insect. Total control will not be achieved using this product alone: 60–80% is the norm. Use when caterpillars are small. Death rates can be lower if caterpillars are mature.

Table 1b: Plant protection products registered for use in an amenity environment – Fungicides

Product	Active ingredient	Comments and pathogens controlled
Systhane	Myclobutanil	Blossom wilt in cherries, apple scab, powdery mildew, rust.
Liquid Copper	Copper oxychloride	For suppression of bacterial diseases such as fireblight. Generally applied as late autumn/winter sprays rather than when the tree is in leaf.
Mancozeb	Mancozeb	Can be used to suppress leaf diseases of amenity trees. Best applied as prevention rather than cure. Mancozeb will have little effect once a tree is infected.
Subdue	Mefenoxam	A systemic fungicide with both preventative and curative activity for the control of *Phytophthora* spp. on ornamentals and trees in amenity situations. Subdue should be applied as a drench to the root of the tree. Subdue should not be used as a foliar spray.
Signum	Pyraclostrobin and Boscalid	A systemic and curative fungicide to control botrytis, fungal leaf spots, powdery mildew, rust, *Venturia* spp. and *Spilocaea* spp., willow anthracnose (*Drepanopeziza sphaeroides*), plane anthracnose (*Apiognomonia*), *Pestalotiopsis*, *Phomopsis*, *Kabatina*, *Meria*, *Gemmamyces*, *Cyclaneusma*, *Lophodermium*, *Sphaeropsis*, *Brunchorstia*, *Ramichloridium* and *Didymascella* (*Keithia*) on amenity vegetation and interior landscapes.

Note: Specific Off Label Approvals (SOLA) are now known as Extension of Authorisation of Mono Use (EAMU) and allow other uses than those shown on the product label.

Approved uses and EAMUs may be checked through the databases held by the Chemicals Regulations Directorate, accessed through the Health and Safety Executive's website: www.hse.gov.uk/presticides.

Reporting pests and diseases

Several agencies in the UK are responsible for collating, monitoring and enforcing control of pests and diseases affecting the nation's tree stock. If you need to report a problem, select the best-fit from the list below. Be aware that you may be directed to one of the other agencies listed as responsibilities are shared with a lead authority in some cases.

Fera's (Food and Environment Research Agency) Plant Health and Seeds Inspectorate (PHSI) is responsible for implementing the plant health regulations in England and Wales on behalf of Defra and the Welsh Government. Fera's PHSI, together with the devolved administrations and the Forestry Commission (FC), forms the UK Plant Health Service and works with other EC Member States and the European Commission to agree appropriate plant health rules for Europe and co-ordinate their implementation.

Fera Plant Health Helpline 0844 2480071 or email planthealth.info@fera.gsi.gov.uk.

FC Plant Health www.forestry.gov.uk/planthealth gives contact details for the Plant Health Service.

Scottish plant health issues are dealt with by the Forestry Commission on behalf of the Scottish administration, see above.

Northern Irish plant health issues are dealt with by: Dept of Agriculture and Rural Development, Dundonald House, Upper Newtownards Road, Ballymiscaw, Belfast BT4 3SB.

Biosecurity

Biosecurity is a set of precautions that aim to prevent the introduction and spread of harmful organisms – pests (e.g. insects), pathogens (e.g. bacteria and fungi) or invasive species (e.g. Japanese knotweed, Oxford ragwort).

Biosecurity measures are the practical steps designed to minimise the risk of introducing or spreading pests and diseases.

Increased global trade and the movement of people and goods between countries mean an increased risk of spreading pests and diseases, which may travel in live plants, in plant products (including seed), in packaging and shipping crates or on vehicles transporting goods or people. Climate change also has the potential to allow some pests and pathogens to travel further and become established in areas where they previously would not survive.

Pests and diseases are not always visible and thus accidental transmission is possible by arboricultural workers and others who routinely work with trees. Pests are most often transported in soil or organic material, such as plant debris, that can be carried on footwear or on the wheels of vehicles.

In terms of practical action, professional tree workers should first be aware of any legal restrictions on the movement of trees. The Arboricultural Association will be active in informing members of these restrictions, which are often issued by government departments such as Defra and their agencies such as the Forestry Commission, Forest Research and Fera. Advice from these agencies may change as a pest outbreak develops so it is essential to be alert.

Current advice on biosecurity is given at www.forestry.gov.uk/biosecurity and involves a risk assessment to separate low-risk from high-risk activities. Keeping footwear, vehicles and equipment clean should be routine and the use of disinfectant may be appropriate where high-risk activities are identified. The issue of personal biosecurity kits may be useful and suitable disinfectant should be available for tools etc. at all times.

Prior to any visit to inspect or work on a site consider the risk involved in terms of pest or disease transmission.	Low Risk Activities – for routine operations with little risk of contact with high risk pest and disease.	High Risk Activities – specialist or targeted operations which may involve contact with infected or infested material or a site that is suspected to harbour the same
Personal Biosecurity Footwear, outerwear and safety equipment	Ensure footwear is capable of being easily cleaned Clean all equipment regularly to remove soil and organic debris.	Plan your visits to include highest risk sites last. Clean footwear and outerwear between each site, using an appropriate disinfectant.
Vehicle Biosecurity Cars, vans, trucks.	Clean vehicles regularly, particularly wheel arches, wheels and tyres to remove mud and organic debris.	Park off-site if possible. Remain on hard surfaced areas. Remove any mud and organic debris on site before leaving. Use an appropriate disinfectant on wheels, tyres and around wheel arches.
Equipment Biosecurity Secateurs, handsaws, chainsaws, chippers, stump grinders etc.	Only take necessary equipment for the task at hand. Remove loose organic debris before leaving site.	Clean and disinfect any tools used during your visit. When taking samples, disinfect equipment between each sample. Keep samples in individually sealed containers.

Personal Biosecurity Kits.

In order to meet reasonable levels of biosecurity the following items should be available for use:

- Clean water (approximately 5 litres);
- Boot tray or bucket big enough to scrub boots in;
- Cleaning equipment e.g. hard brush, tread scraper, portable sprayer;
- Disinfectant in sealed container;
- Personal PPE for application, disposable overall, gloves, eye protection;
- Cleansing wipes, paper towels;
- Sealable bags for contaminated equipment.

Disinfectants

Alcohol based disinfectants (for example industrial methylated spirit or isopropyl) at 70% concentrations are recommended as being effective against Phytophthora and other pathogens. All such disinfectants fall under the provisions of the Control of Substances Hazardous to Health Regulations 2002. When using disinfectants always:

- Read and follow the manufacturer's instructions on the product label;
- Correctly use any required PPE;
- Mix and use in a well ventilated area;
- Do not use where run off may enter watercourses, surface water drains, springs or wells;
- Remove loose debris first and apply to clean surfaces;
- Ensure any contact time and wash off is observed.

Index of scientific names

	Page
Anoplophora glabripennis (Asian longhorn beetle)	6
Apiognomonia erythrostoma (cherry leaf scorch)	8
Apiognomonia veneta (anthracnose of London plane)	10
Armillaria mellea/solidipes (=ostoyae) (honey fungus)	12
Blumeriella jaapii (cherry leaf spot)	8
Cameraria ohridella (horse chestnut leaf miner)	14
Chalara fraxinea (ash dieback)	18
Cinara cupressi (cypress aphid)	20
Cossus cossus (goat moth)	22
Cristulariella depraedans (a leaf disease of maples)	60
Cryptococcus fagisuga (felted beech scale)	24
Cryptostroma corticale (sooty bark disease)	26
Dendrolimus pini (pine tree lappet moth)	28
Dothistroma septosporum (red band needle blight)	30
Drepanopeziza sphaeroides (anthracnose of willow)	32
Erwinia amylovora (fireblight)	34
Erysiphe alphitoides Oak leaf powdery milldew	36
Euproctis chrysorrhoea (brown tail moth)	38
Guignardia aesculi (horse chestnut leaf blotch)	40
Lymantria dispar dispar (European gypsy moth)	44
Melampsora or *Melampsoridum* spp. (rusts of poplar, birch and willow)	46
Monilinia laxa (blossom wilt and wither tip)	8
Neonectria coccinea (beech bark disease)	48
Ophiostoma novo-ulmi (Dutch elm disease)	52
Phytophthora species	54
Pseudomonas syringae pv. *aesculi* (bleeding canker of horse chestnut)	56
Pulvinaria regalis (horse chestnut scale)	58
Rhytisma acerinum (tar spot of sycamore)	60
Seridium cardinal (Coryneum canker)	62
Sesia apiformis (hornet moth)	64
Splanchnonema platani (Massaria)	66
Taphrina spp. (witches' broom)	68
Thaumetopoea processionea (oak processionary moth)	70
Venturia spp. (scab)	72
Zeuzera pyrina (leopard moth)	74